RAND NATIONAL DEFENSE RESEARCH INSTITUTE

Defense Working Capital Fund Pricing in the Defense Finance and Accounting Service

A Useful, but Limited, Tool

Edward G. Keating, Ellen M. Pint, Christina Panis, Michael H. Powell, Sarah H. Bana

Prepared for the Office of the Secretary of Defense

For more information on this publication, visit www.rand.org/t/RR866

Library of Congress Cataloging-in-Publication Data is available for this publication.
ISBN: 978-0-8330-8869-7

Published by the RAND Corporation, Santa Monica, Calif.

Preface

During the early fall of 2013, the Office of the Secretary of Defense for Cost Assessment and Program Evaluation (OSD-CAPE) asked the RAND Corporation to assess the advantages and disadvantages of continuing to fund the Defense Finance and Accounting Service (DFAS) through a Defense Working Capital Fund (DWCF).

Abetted by reduced prices made possible by using automated approaches, DFAS has successfully induced its clients to evolve toward less costly approaches for paying Department of Defense (DoD) contractors and personnel.

DWCF prices provide more incentives to DFAS customers than to DFAS itself. However, DFAS's constant dollar costs have fallen over time, even as overall DoD spending has increased.

On balance, we do not recommend that DFAS return to being funded solely by direct appropriation. However, it may be beneficial to reform DFAS (and, more generally, DWCF) pricing to allow nonlinear approaches, such as quantity discounts and direct funding of fixed costs.

This research report should be of interest to DoD personnel involved with DWCF and transfer pricing issues. It was sponsored by OSD-CAPE and conducted within the Acquisition and Technology Policy Center of the RAND National Defense Research Institute, a federally funded research and development center sponsored by the Office of the Secretary of Defense, the Joint Staff, the United Combatant Commands, the Navy, the Marine Corps, the defense agencies, and the defense Intelligence Community.

For more information on the RAND Acquisition and Technology Policy Center, see http://www.rand.org/nsrd/ndri/centers/atp.html or contact the director (contact information is provided on the web page).

Contents

Figures and Table

Figures

Table

Summary

Building on previous RAND research on Defense Working Capital Fund (DWCF) pricing policies and the Defense Finance and Accounting Service (DFAS), the Office of the Secretary of Defense for Cost Assessment and Program Evaluation (OSD-CAPE) asked RAND to undertake a research project on understanding the advantages and disadvantages of funding DFAS through a DWCF versus using direct appropriations. As a DWCF entity, DFAS covers its costs by selling its services to customers, primarily within the Department of Defense (DoD). More broadly, the DWCF approach is intended to create businesslike incentives for both customers and service providers to reduce costs.

Under DoD policy, DWCF providers have historically used what might be termed *expected average cost* pricing. In this case, a DWCF provider, in conjunction with the customer, estimates how much and what types of work the DWCF provider will perform, along with the costs it will incur. Costs are allocated across the provider's products. Then, each product's price is set as the ratio of expected costs allocated to the product divided by the quantity expected to be sold. Through this approach, the intent is for the DWCF provider to break even.

In evaluating the desirability of DWCF funding of DFAS, we analyze two broad categories of criteria by asking two questions:

1. Does the DWCF structure provide appropriate incentives to DFAS?
2. Does the DWCF structure provide appropriate incentives to DFAS's customers?

To address these questions, we used two complementary methodologies: We analyzed DFAS cost and workload data, and we conducted a series of interviews with subject-matter experts.

An Overview of DFAS Operations

DFAS provides finance and accounting services to DoD (and a few non-DoD) customers. In DFAS's vernacular, *finance* refers to the act of paying someone, such as a member of the military or a defense contractor. *Accounting* refers to the agglomeration and analysis of accounting data.

DFAS was formed in 1991, consolidating what had been service-specific operations. The intent was to reduce the cost of DoD finance and accounting operations while strengthening its financial management. DFAS's real costs have been trending downward since the formation of the agency. Real costs in 2014 were 58 percent of real costs in 2000. DFAS's civilian end strength declined from 17,344 in fiscal year (FY) 2000 to an estimated 12,014 in FY 2014.

Only about half of DFAS costs are labeled as *direct* (e.g., assigned to specific outputs), with the plurality of the remaining costs categorized as *indirect*. DFAS's direct costs divide into three broad categories: accounting services, finance services, and direct costs that are billed to clients but not attributed to specific outputs.

Insights from DFAS Cost-Workload Data

DFAS has used reduced prices across several outputs to encourage customers to adopt approaches that are more automated and less expensive. For example, customers have transitioned the majority of their commercial payments from manual to electronic formats, at least partially in response to lower prices for electronic commerce. However, other types of workload, such as military retired pay accounts, are almost completely unresponsive to price changes, because customers cannot adjust the number of retired military personnel when prices change.

DFAS has also implemented customer-specific pricing for several outputs, including active-duty military pay accounts, thereby rewarding customers who put fewer burdens on DFAS. However, the cost data that we received from DFAS did not include customer-level attribution (i.e., which costs were borne in support of which customers), so we were unable to assess the validity of DFAS's customer-specific prices.

Assessing DWCF Pricing in DFAS

There are both advantages and disadvantages of DWCF pricing in DFAS.

Advantages

DFAS has had success transitioning customers to more-automated approaches. DWCF pricing encourages such evolution by rewarding customers with lower prices when they switch to automated approaches. However, there were concurrent regulatory pressures to adopt such approaches, so it is hard to know to what extent reduced prices contributed to the favorable outcome. DFAS's customer-specific pricing does have the favorable implication of rewarding customers who put fewer burdens on DFAS and encouraging other customers to do likewise (though RAND lacked visibility into customer-attributed costs to validate these prices). DWCF pricing encourages ongoing cost-related dialog both within DFAS and between DFAS and its customers.

Also, being a DWCF entity provides greater managerial flexibility to DFAS than if it were solely dependent on direct appropriations. For example, it can meet increased demands without requesting additional appropriations and can continue to operate during a budget sequester. In addition, the flexibility allows DFAS to take on non-DoD workload, which it otherwise could not do. Limited evidence suggests that non-DoD customers more than pay their way—that is, they reduce the amount of DFAS overhead that must be covered by DoD customers—further benefiting DoD.

Disadvantages

The structure appears to impose a greater cost accounting burden than would a direct appropriation mechanism, because DFAS has to attribute its costs to specific products.

We are also concerned that customers may be developing skepticism about the actual consequences of their cost-saving steps. We heard customers say that DFAS will simply raise prices and rates on other workload even if it provides a reduced price on a specific output. DFAS only saves marginal costs when customers evolve toward more-automated approaches, so fixed costs must be re-allocated to remaining workload.

DFAS's DWCF expected average cost prices are almost certainly above marginal costs, which means that DFAS costs will not fall commensurably when workload falls. This phenomenon is a source of recurring disappointment for DFAS customers, whose bills do not fall as much as expected in response to customer cost-saving reforms.

Conclusions and Recommendations

Because DFAS is a monopoly, it may lack robust incentives to provide its services in a cost-effective way. However, DFAS's costs have trended down without the agency facing the spur of competition to reduce its costs. Customer and DoD oversight pressure, as well as the intrinsic motivation of DFAS leaders, have sufficed, at least heretofore, to achieve a favorable cost trend.

If DFAS were funded directly by appropriations, DFAS could be told exactly the budget level at which it must operate. However, it might then be unable to fulfill its mission satisfactorily.

An aggressive reform that would change DFAS's incentive structure considerably would be to modify DoD policy to allow other governmental or private-sector providers to compete with DFAS for DoD business. In other parts of DoD, DWCF pricing rules have caused problems when customers have had alternatives. Assessing the overall desirability of allowing competition with DFAS was beyond the scope of this analysis.

DWCF prices provide considerable incentives to DFAS customers. Customers are rewarded for adopting approaches that put less burden on DFAS. However, because prices include both fixed and variable costs, DFAS's actual costs do not fall as much as the DWCF price reductions that customers receive. As a result, customers have been disappointed that their DFAS bills did not fall as much as they expected when they implemented more-automated approaches.

Nonlinear pricing (e.g., quantity discounts, DWCF provider fixed costs being funded by appropriation) would address concerns with inapt customer incentives and expectations.

We urge pilot project experiments to better understand nonlinear pricing's implementation challenges, costs, and ultimate consequences.

Acknowledgments

Michael Strobl of OSD-CAPE served ably as this project's client point of contact. We also appreciate assistance from his colleague Andrew Mara and input and suggestions from Rick Burke, Scott Comes, Veronica Daigle, Preston Dunlap, Danielle Miller, and Jerry Pannullo of OSD-CAPE.

Dan Walsh was our point of contact within DFAS, and we would also like to thank Paul Gass of DFAS for orchestrating some of our subject-matter expert interviews. We also appreciate assistance from Steve Burghardt of DFAS.

We appreciate the time and insights that we received from our subject-matter experts, including Gretchen Anderson (Office of the Secretary of Defense, director for revolving funds), Gregory Bitz (U.S. Navy and former director of DFAS's Kansas City and Indianapolis centers), Bruce M. Carnes (former DFAS chief financial officer and deputy director), Eric Cuebas (U.S. Air Force), Zack Gaddy (former DFAS director), Ray Gaw (DFAS Indianapolis), Stephen Herrera (U.S. Air Force), Tony Hullinger (DFAS Indianapolis), Cynthia Jones (Office of the Secretary of Defense [Comptroller]), Rich Luster (DFAS Indianapolis), Eric Reid (U.S. Army), Dennis Taitano (U.S. Navy), Erica Thomas (U.S. Navy), and James Watkins (U.S. Army).

Cynthia R. Cook and Marc Robbins provided program leadership to the RAND research team. We appreciate their efforts on our behalf. Dr. Robbins provided numerous helpful suggestions as this research evolved. Allison Kerns edited this report. We also appreciate assistance from our RAND colleagues Eric Peltz and Guy Weichenberg.

We received thoughtful and constructive reviews of an earlier version of this report from Dr. Carnes and from our RAND colleague Susan M. Gates.

Abbreviations

BRAC	Base Realignment and Closure
DFARS	Defense Federal Acquisition Regulation Supplement
DFAS	Defense Finance and Accounting Service
DoD	U.S. Department of Defense
DWCF	Defense Working Capital Fund
FORSCOM	Forces Command (U.S. Army)
FY	fiscal year
G&A	general and administrative
LES	leave-and-earning statement
OSD-CAPE	Office of the Secretary of Defense for Cost Assessment and Program Evaluation
SME	subject-matter expert

CHAPTER ONE

Introduction

More than ten years ago, the management of the Defense Finance and Accounting Service (DFAS) hired the RAND Corporation to undertake a series of analyses, resulting in three separate reports that discussed the management and operation of DFAS (Keating and Gates, 1999; Keating et al., 2001; and Keating et al., 2003). A recurring topic in those reports was Defense Working Capital Fund (DWCF) pricing rules and the extent to which DWCF rules interfaced imperfectly with DFAS's cost structure, a cost structure with considerable fixed (output-invariant) costs.[1] RAND used DFAS cost and workload data to argue for broader DWCF reform, such as allowing DWCF providers to use nonlinear pricing or using appropriations to cover DWCF providers' fixed costs (Keating, 2001).

Canonically, a government organization is funded through appropriations. For this sort of system, a budget is passed specifying that a certain amount of funding is to be devoted to a project or function, funds are provided, and the project's or function's managers are responsible for achieving what is desired with those funds. If desired missions increase or costs are found to be greater than expected, additional appropriations may be needed.

The DWCF approach is different: It makes a governmental organization at least partially dependent on payments from other govern-

[1] The nomenclature *fixed cost* does not mean that a cost cannot be cut. Rather, the term refers to cost levels that are unrelated, at least in the short run, to the firm's level of production or sales. For example, a firm's expenditures on corporate management are unlikely to be changed based on a short-run change in sales levels.

mental organizations. A DWCF provider no longer receives full (or perhaps even any) appropriations. Instead, the provider must raise revenue from other organizations by selling them goods and services. Those revenue-providing organizations must, either directly or indirectly, be receiving appropriations, but the DWCF provider is "downstream" from those appropriations.[2]

DWCF providers have historically used what might be termed *expected average cost* pricing.[3] In this case, a DWCF provider, in conjunction with the customer, estimates how much and what types of work the DWCF provider will perform, along with the costs it will incur. Costs are allocated across the provider's products. Then, each product's price is set as the ratio of expected costs allocated to the product divided by the quantity expected to be sold. The intent is for the DWCF provider to break even. If workload levels and costs occur as expected, DWCF provider revenue will equal its costs.

The DWCF price-setting process is lengthy. Cost and workload estimates are typically generated two years in advance to help appropriation-dependent customers generate their budgets. But if conditions then change (e.g., more or less workload than anticipated; a change in input costs, such as energy), the DWCF provider is stuck with an annual stabilized rate structure that is fated to generate unintended profits or losses. Realized profits or losses, although clearly sunk, are then to be rebated (for profits) or recovered (for losses, the more common case) from future customers, but those rebates or surcharges do not affect prices until several years later, when they have been worked through the lengthy budget process.[4] Today's customer may benefit from or be penalized by circumstances that affected potentially different customers several years ago.

---

[2] Byrnes (1993) provides a description of the DWCF approach, though using the since-replaced terminology *Defense Business Operations Fund.*

[3] U.S. General Accounting Office (1997) describes the DWCF price-setting process.

[4] Friend (1995) discusses a Navy aviation example in which unplanned engine work caused losses that resulted in a sharp increase in future stabilized rates. In that example, the Navy requested a pass-through—that is, appropriation—from Congress to keep those losses from raising future rates unduly.

According to the U.S. Department of Defense (DoD) Financial Management Regulation and 10 U.S.C. 2208, most DWCF activities are required to set their prices based on full cost recovery, including all general and administrative support provided by others. Prices that are established through the budget process must remain fixed during the year of execution, except for unusual circumstances. In addition, prices must be set to break even in the long run—that is, to make up actual or projected losses or to return actual or projected gains (DoD, 2010b, p. 9-20). However, the regulation does allow DWCF activities to receive direct appropriations for two general purposes: to provide working capital (e.g., when the cumulative operating results or the cash position is negative) and to provide financing for specific projects or tasks (e.g., excess capacity to respond to contingencies at industrial activities, the costs of emergency or humanitarian missions at U.S. Transportation Command, and operating costs at the Defense Commissary Agency) (DoD, 2013b, pp. 1-7 and 3-3).

DFAS has changed considerably since RAND's earlier analyses (Keating and Gates, 1999; Keating et al., 2001; and Keating et al., 2003). Most pronouncedly, several DFAS facilities were closed as a result of the 2005 round of Base Realignment and Closure (BRAC) decisions. As we expand upon in Chapter Two, today's DFAS is a more geographically concentrated organization with a lower level of staffing and annual costs (in constant dollars) than it was when we undertook our previous analyses.

But DFAS's role remains broadly the same: providing finance and accounting services to DoD and a small number of non-DoD customers. DFAS does not receive any direct appropriations; it is "downstream" from customers who receive appropriations (directly, such as the military services, or indirectly, such as the Defense Logistics Agency, itself a DWCF provider). DFAS uses customer-specific pricing more now than was true when we undertook our earlier analyses.

In the summer of 2013, the Office of the Secretary of Defense for Cost Assessment and Program Evaluation (OSD-CAPE) asked RAND to research the advantages and disadvantages of funding DFAS through a DWCF versus using direct appropriations.

In evaluating the desirability of DWCF funding of DFAS, we analyze two broad categories of criteria by asking two questions:

1. Does the DWCF structure most effectively provide incentives to DFAS?
2. Does the DWCF structure most effectively provide incentives to DFAS's customers?

The answers to these questions then affect whether the DWCF structure allows DFAS to provide requisite services to DoD customers in a maximally cost-effective manner.

We used two complementary methodologies to address the questions. First, we analyzed data on DFAS costs (how much DFAS has spent in various categories to provide services) and workload (the quantities and types of services that DFAS has provided to its customers). This analysis's primary cost-workload data set covered fiscal years (FYs) 2005 through 2013. We also had access to DFAS's historical price lists.

Second, the RAND research team conducted a series of subject-matter expert (SME) interviews. Our interviewees fell into four broad categories: current DFAS employees, former DFAS employees, DFAS customers, and Office of the Secretary of Defense (Comptroller) employees. These roughly one dozen interviews occurred between December 2013 and May 2014. We asked these SMEs about their experiences with DFAS, their perceptions of its strengths and weaknesses, and their views on the desirability and applicability of DWCF-type pricing to DFAS. While we enumerate our interviewees in the acknowledgments, we will not associate specific views to specific individuals. We are very grateful for the candid and insightful comments that we received.

The remainder of this report is structured as follows: Chapter Two provides an overview of DFAS operations. Chapter Three uses the FY 2005 through FY 2013 DFAS cost-workload data to provide insight into DFAS's cost structure, such as the relationship between how much work DFAS performs and how much DFAS operations cost. Chapter Four then uses these cost-workload insights to assess the applicability and desirability of DWCF pricing to DFAS. In Chapter Five, we summarize our findings on the DWCF structure's effects on incentives to DFAS and its customers. The appendix provides the question protocols that we used in our SME interviews.

CHAPTER TWO

An Overview of DFAS Operations

As the name suggests, DFAS provides finance and accounting services to DoD (and a few non-DoD) customers. In DFAS's vernacular, *finance* refers to the act of paying someone—for example, members of the military, government-employed civilians, and defense contractors. *Accounting* refers to the agglomeration and analysis of accounting data for both financial accounting (mandated reporting) and managerial accounting (leadership decisionmaking) purposes.

Beyond the direct actions of paying people and preparing accounting reports, DFAS has a responsibility to develop finance and accounting systems to replace obsolete systems, to consolidate disparate legacy systems (a responsibility that is diminishing over time), and to modify systems to comply with changes in Federal Accounting Standards Board rules, federal regulations, and legislative mandates. These system management tasks do not vary directly with DFAS's day-to-day workload, so DFAS's per-output costs tend to increase when DFAS's workload declines. Even if the costs of system management are attributed to specific outputs, these costs do not decline when workload declines, resulting in average cost increasing when workload falls.

DFAS was created in 1991, consolidating what had been finance and accounting operations specific to military services. The intent was to reduce the cost of these operations for DoD while strengthening its financial management. Since its inception, DFAS has reduced the number of different finance and accounting systems in use from 330 to 111 (DFAS, 2014).

While the service-specific operations that were combined to form DFAS encompassed more than 300 installation-level offices, DFAS operations today are concentrated in five main operational sites: Cleveland, Ohio; Columbus, Ohio; Indianapolis, Indiana; Limestone, Maine; and Rome, New York. The current structure came to be after the 2005 BRAC process that closed 20 DFAS facilities (DFAS, 2007). Table 2.1 enumerates the customers served and outputs provided by the five main operational sites today.

Data from the FY 2000–2014 DFAS budget submissions show that DFAS's civilian end strength declined from 17,344 in FY 2000 to an estimated 12,014 in FY 2014. (In Figures 2.1 and 2.2, FY 2000–2012 data are actuals, while FY 2013 and FY 2014 values are budget

Table 2.1
DFAS Main Operational Sites, Customers, and Outputs

Location	Customers	Outputs
Cleveland, Ohio	Marine Corps Navy	Civilian pay Military pay Retired and annuity pay Disbursing Accounts maintenance and control Accounts payable Accounts receivable
Columbus, Ohio	Defense Logistics Agency Air Force Army Navy Marine Corps	Disbursing Accounts maintenance and control Accounts payable Accounts receivable
Indianapolis, Indiana	Air Force Army Navy Marine Corps	Civilian pay Military pay Travel pay Transportation pay Disbursing Accounts maintenance and control Accounts payable Accounts receivable
Limestone, Maine	Air Force	Transportation pay Accounts maintenance and control Accounts payable Accounts receivable
Rome, New York	Army	Travel pay Accounts maintenance and control Accounts payable Accounts receivable

SOURCE: DFAS, 2011.

Figure 2.1
DFAS Staffing Levels, Fiscal Year 2000–2014

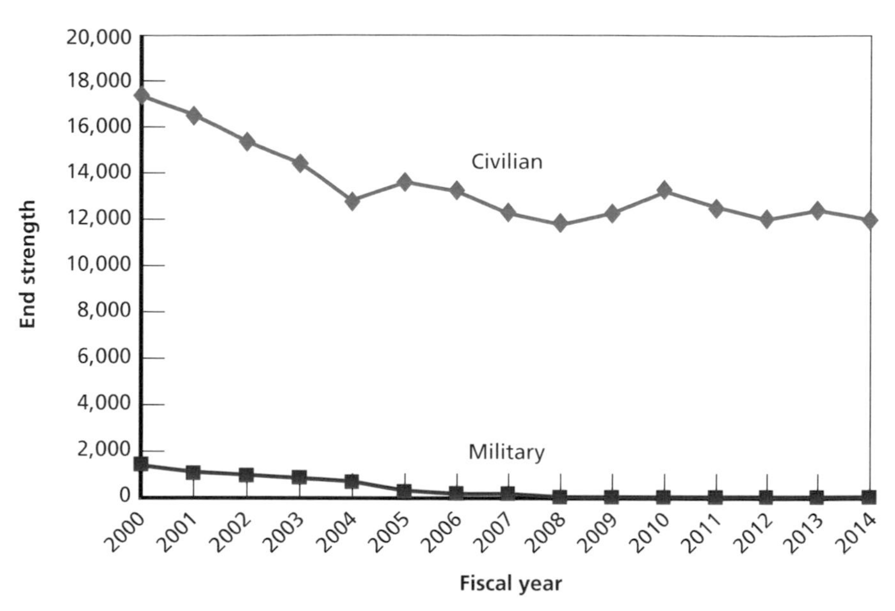

SOURCE: DoD budget documentation and budget estimates (DoD, various years).
RAND *RR866-2.1*

estimates.) Meanwhile, military staffing in DFAS has all but disappeared, plunging from 1,428 in FY 2000 to 29 in FY 2014.

Much of the civilian personnel decline occurred between FY 2000 (17,344) and FY 2004 (12,826). DFAS civilian staffing has been more static since then. Notice that the decline largely preceded the 2005 BRAC round. In Keating and Gates (1999), we showed a general pattern of decline in DFAS regions' civilian work years from calendar years 1996 to 1998, so Figure 2.1's trend commenced prior to the period presented there.

As shown in Figure 2.2, DFAS's real costs have been trending downward for many years. Real costs in FY 2014 were 58 percent of real costs in FY 2000. That trend also preceded the 2005 BRAC round.

To clarify our vernacular, *cost* refers to costs incurred by DFAS in providing services to its customers and is distinct from what DFAS charges customers for those services (i.e., DFAS's prices or rates).

Figure 2.2
DFAS Constant Dollar Total Costs, Fiscal Year 2000–2014

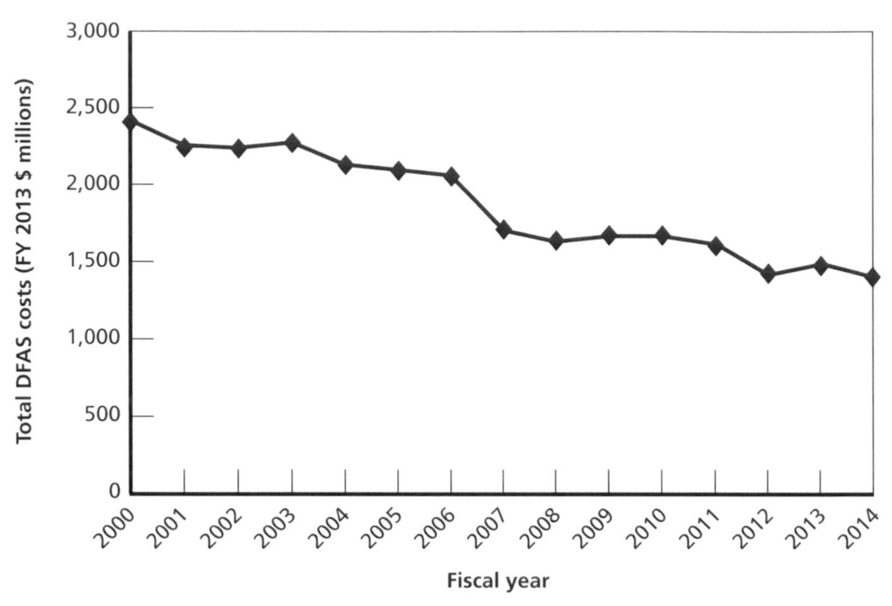

SOURCE: DoD budget documentation and budget estimates (DoD, various years).
RAND *RR866-2.2*

What customers see as the costs of DFAS are revenues from DFAS's perspective.

While DFAS constant dollar total costs have declined since 2000, Figure 2.3 shows that the number of DoD-employed civilian personnel and military personnel has increased. It is reasonable to think that having more DoD employees would, other things being equal, increase burden on DFAS. Also, both the overall DoD budget and the DoD procurement budget increased markedly in real terms to cover the costs of operations in Iraq and Afghanistan.

DFAS deserves credit for the favorable story presented in Figure 2.3: While the size of the DoD—as measured by civilian and military personnel levels, as well as by budgets—has increased since 2000, DFAS's constant dollar total costs have declined considerably.

Figure 2.3
DFAS Constant Dollar Total Costs Compared with DoD Benchmarks, Fiscal Year 2000–2014

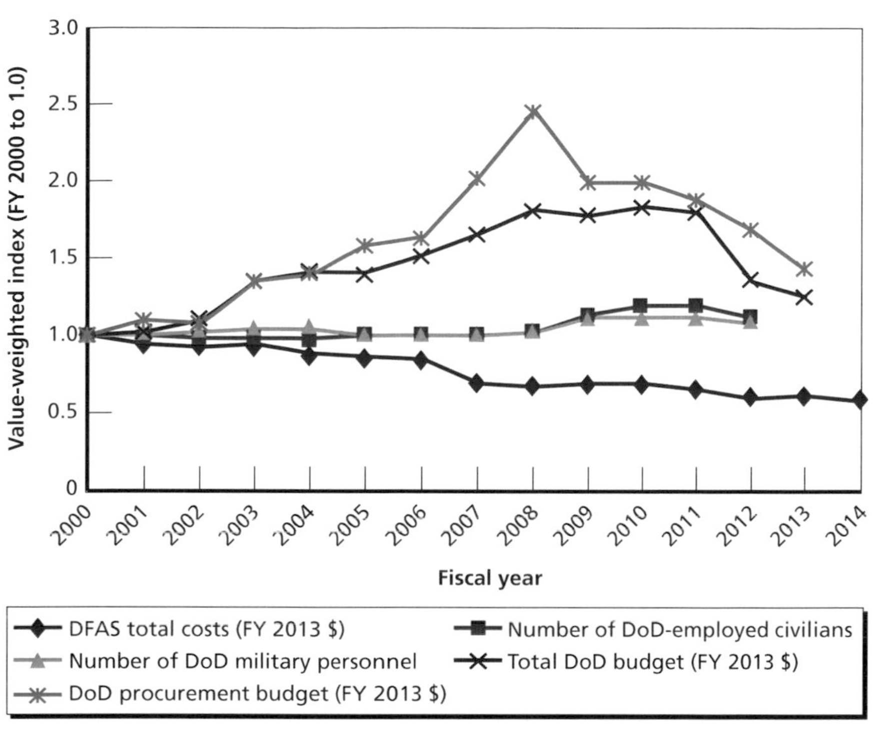

SOURCE: DoD budget documentation and budget estimates (DoD, various years); Office of Personnel Management, undated a, b.
RAND *RR866-2.3*

Not surprisingly, as shown in Figure 2.4, civilian personnel costs have dominated DFAS expenses. DFAS's civilian personnel costs in FY 2014 were about 81 percent of what they were in FY 2000 (in constant dollars). Several other cost categories have had much steeper declines. For example, FY 2014 costs for other purchased services were 50 percent of the FY 2000 level, and FY 2014 costs for capital depreciation were 16 percent of the FY 2000 level. While DFAS civilian personnel costs have declined, the steeper declines in other cost categories have increased the civilian personnel share of total costs from 51 percent in FY 2000 to 71 percent in FY 2014.

Figure 2.4
DFAS Constant Dollar Costs, by Cost Category, Fiscal Year 2000–2014

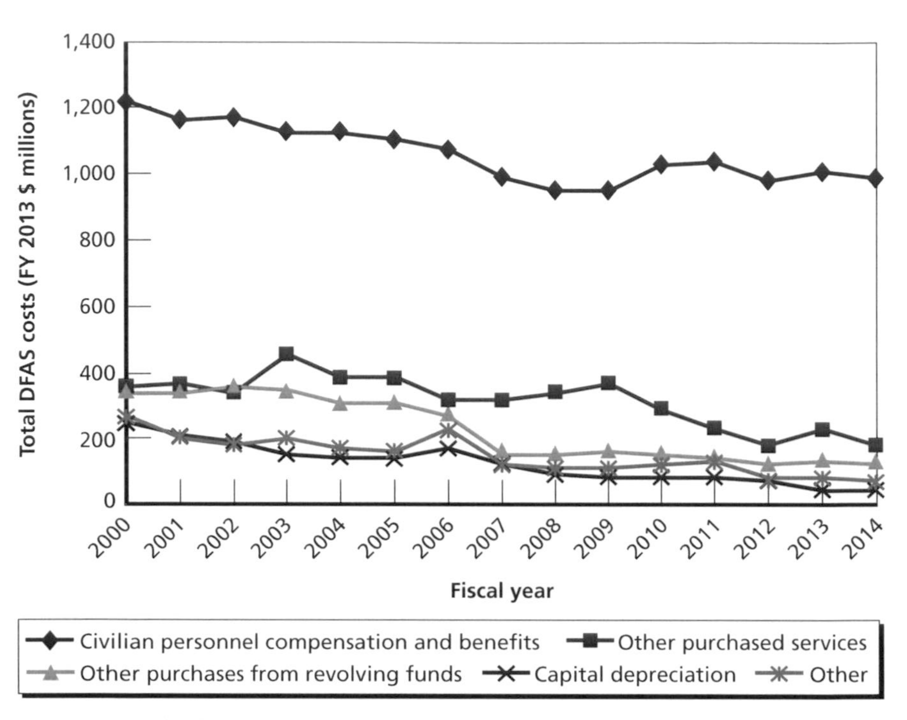

SOURCE: DoD budget documentation and budget estimates (DoD, various years).
RAND *RR866-2.4*

Other purchased services include facility rental and maintenance costs. *Other purchases from revolving funds* include, for instance, DFAS purchases of computer support from the Defense Information Systems Agency.

The FY 2005–2013 cost data that RAND received from DFAS provide us with more insight into the nature of the costs (albeit over fewer years). Figure 2.5 shows that only about half of DFAS costs are labeled as *direct* (i.e., assigned to specific outputs), with the plurality of the remaining costs categorized as *indirect.*

DFAS personnel said that their indirect costs, such as supervisory personnel (who do not generate direct billable hours) and automated systems, can still be attributed to workload in the price formulation

Figure 2.5
DFAS Constant Dollar Costs, by Direct and Indirect Categories, Fiscal Year 2005–2013

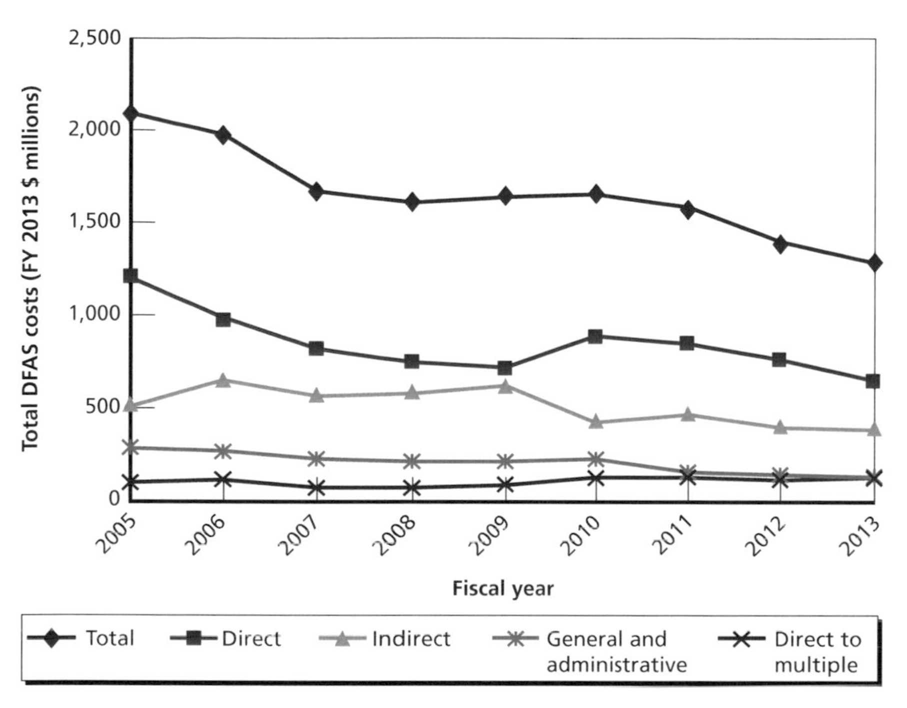

SOURCE: DFAS-provided annual cost data.
RAND *RR866-2.5*

process. Only general and administrative (G&A) costs cannot be specifically associated with DFAS products or outputs. However, we did not receive information on how DFAS assigns indirect costs or the costs of what it terms *direct to multiple* [outputs]. DFAS experts told us that only G&A costs are not assigned to specific outputs; instead, they are allocated in a single percentage across all outputs.

The total annual dollar values of the DFAS-provided data used in Figure 2.5 and the DFAS budget submission data used in Figures 2.2–2.4 are close, though not identical.

As shown in Figure 2.6, DFAS's direct costs divide into three broad categories. Accounting services make up the single most costly type of DFAS output. Second, there are direct costs that are directly billed to clients but not attributed to specific outputs. Third, DFAS has a variety of finance-related outputs, of which active military pay and commercial payments have been most costly.

The DFAS budget submission revenue data presented in Figure 2.7 show that the Army has long been DFAS's largest customer. The Army's share of DFAS's total revenue has trended upward in the past decade. The Air Force and Navy have alternated between being DFAS's second- and third-highest revenue customers, with their respective shares of DFAS's total revenue showing a modest downward trend.

Note that DFAS needs a pricing system in order to take on non-DoD workload. As we discuss in Chapter Four, non-DoD work appears to have helped cover DFAS overhead costs that would have otherwise been entirely borne by DoD customers. Complementing the budget submission revenue data, we combined DFAS prices

Figure 2.6
Output Shares of DFAS Direct Costs, Fiscal Year 2005–2013

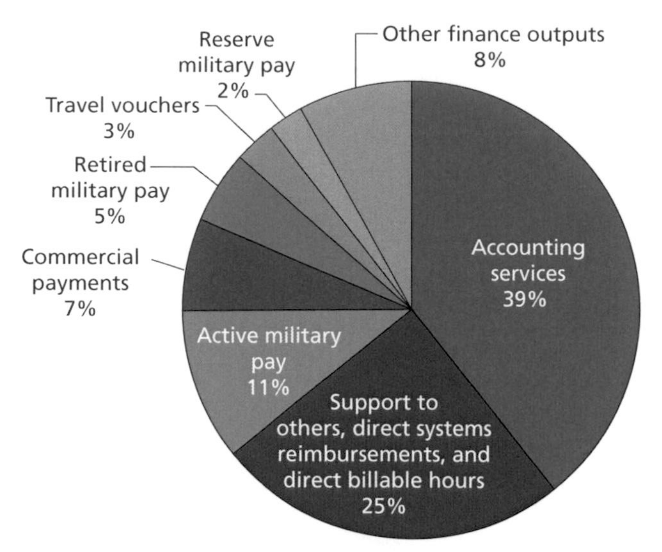

SOURCE: DFAS-provided annual cost data.
RAND *RR866-2.6*

Figure 2.7
Customer Shares of DFAS Revenue, Fiscal Year 2000–2014

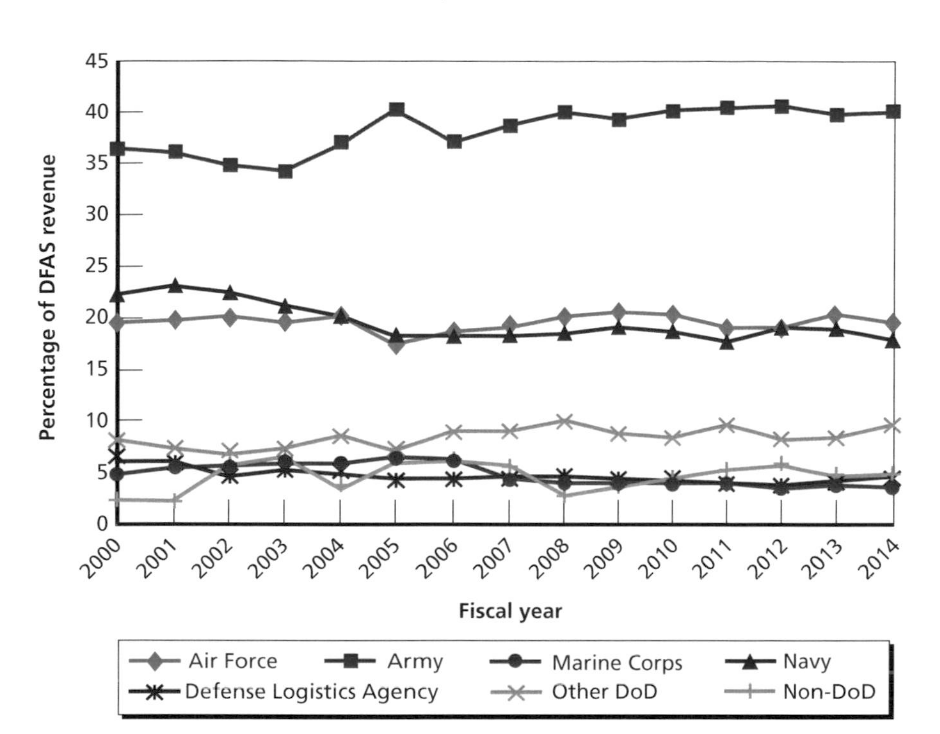

SOURCE: DoD budget documentation and budget estimates (DoD, various years).
RAND *RR866-2.7*

with the workload data to estimate DFAS revenue generated through charges per unit of work. For example, in FY 2013, DFAS provided the Army with about 2.3 million hours of accounting service labor at $68.30 per hour, for a total Army accounting bill from DFAS of about $160 million.

However, as shown in Figure 2.8, FY 2005–2013 DFAS budget submissions showed greater total DFAS revenue than we calculated from the workload data aggregating across all DFAS customers and outputs. Discussions with DFAS personnel confirmed that DFAS receives additional revenue from customers outside of charges per unit of work.

In the next chapter, we draw insights from DFAS cost-workload data.

Figure 2.8
Not All DFAS Revenue Comes from Charges Per Unit of Work, Fiscal Year 2005–2013

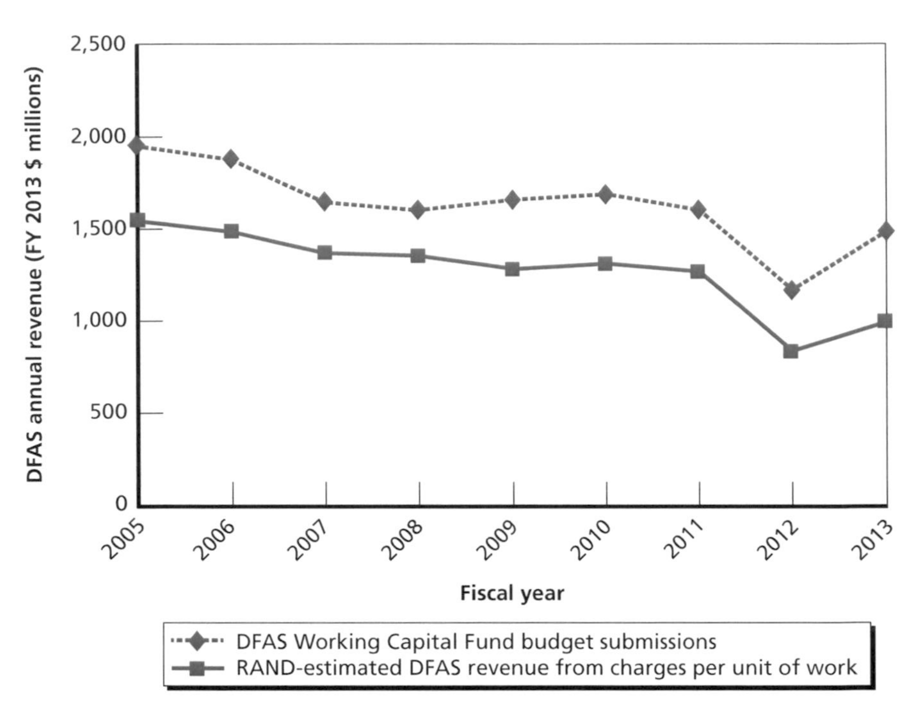

SOURCE: DoD budget documentation and budget estimates (DoD, various years); DFAS-provided annual workload data.

RAND *RR866-2.8*

CHAPTER THREE

Insights from DFAS Cost-Workload Data

Perhaps the most persuasive argument in favor of DWCF pricing in DFAS is that it provides an opportunity for DFAS to incentivize customers to adopt more-automated, lower-cost approaches.

DFAS endeavors to align pricing for services with the costs of providing the service, to the extent possible. Thus, when different methods of service have different cost structures, it uses differentiated pricing. Reflecting automated outputs' reduced costs to DFAS, DFAS offers lower prices for using automated approaches than it does for manual approaches across several categories or families of outputs, with further differentiation among automated approaches when applicable. Illustrating this phenomenon, Figure 3.1 shows the prices faced by the Army for Output 09 (commercial payments), Output 29 (commercial payments—electronic commerce), and Output 49 (fully electronic vendor pay).[1] Output 09 is the most manual and most expensive of the three approaches to making commercial payments, and Output 49 is the most automated and least costly. While prices have varied considerably year-to-year (with increasing Output 29 prices from FYs 2007 to 2011), the more-automated approaches have consistently had lower prices. Irrespective of the year of transition, customers were rewarded for transitioning from Output 09 to Output 29 and from Output 29 to Output 49.

Because DFAS customers have transitioned to more-automated approaches for commercial payments, such lower prices for these

[1] The Army is DFAS's largest customer, and the pattern of commercial payment prices faced by the Army is representative.

Figure 3.1
DFAS Prices for the Army Over Time, by Type of Commercial Payment, Fiscal Year 2005–2013

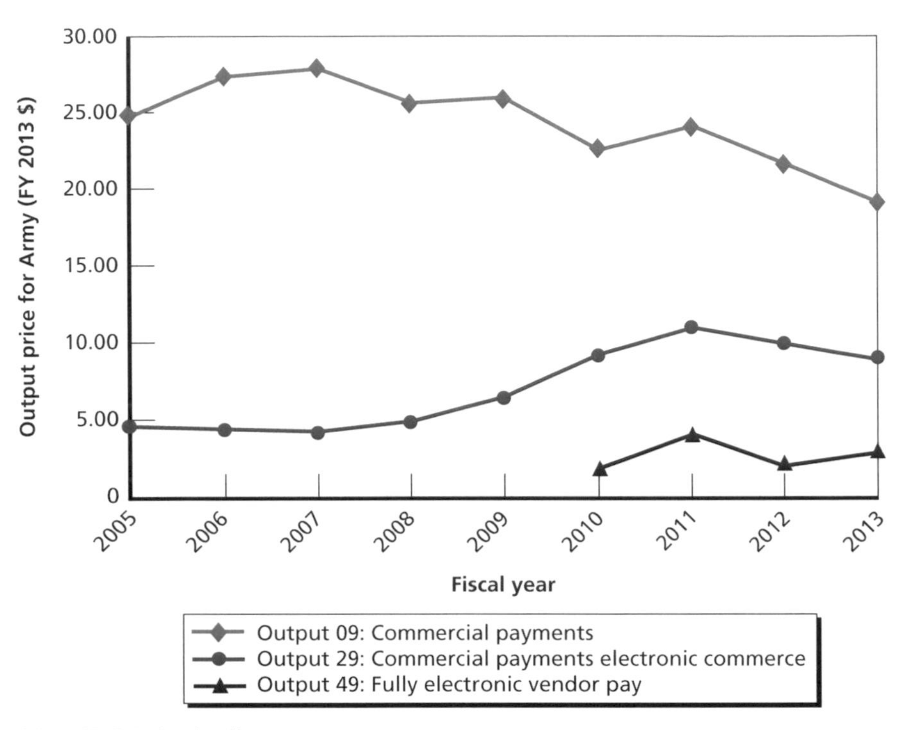

SOURCE: DFAS price lists.
RAND *RR866-3.1*

approaches (Output 29's price always being lower than Output 09's, and Output 49's price always being lower than Output 29's) have seemingly been effective incentives, as shown in Figure 3.2.

In 2013, the most automated approach to making commercial payments, Output 49, became the most widely used. We caution, however, against giving all of the credit for the recent adoption of automated commercial payments to DFAS's price incentives. Concurrent to lower prices for automated approaches, a variety of mandates have been imposed to drive DoD customers toward such approaches. For example, in 2004, the Defense Federal Acquisition Regulation Supplement 252.232-7003, "Electronic Submission of Payment

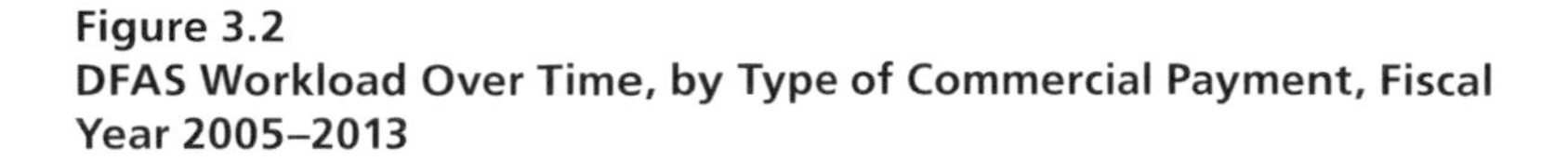

Figure 3.2
DFAS Workload Over Time, by Type of Commercial Payment, Fiscal Year 2005–2013

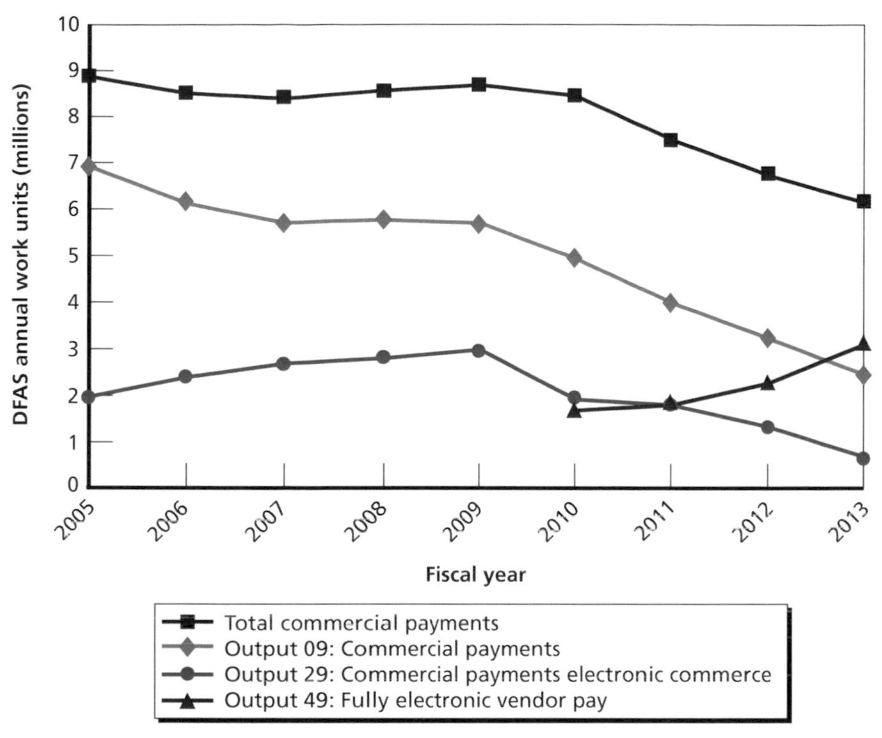

SOURCE: DFAS-provided annual workload data.
RAND *RR866-3.2*

Requests and Receiving Reports," required claims for payment under a DoD contract to be submitted in electronic form (*Federal Register*, 2012). Also, in 2007, the DoD proposed to amend that supplement to require use of the Wide Area WorkFlow Receipt and Acceptance electronic system for submitting and processing payment requests under DoD contracts (*Federal Register*, 2007). DFAS pricing has been only one contributor to the observed evolution toward more-automated approaches. Customer interviewees were aware of these reduced prices and agreed that they contributed to the observed shifts in workload toward automated approaches.

Figure 3.3 shows a similar pattern for *leave-and-earning statements* (LESs), the statements that workers receive with their pay checks, with the lower-priced, more-automated approach gaining market share over time.

As was true with commercial payment automation, there were concurrent complementary policy changes driving DoD customers toward automated approaches. For example, on April 29, 2005, the Under Secretary of Defense for Personnel and Readiness and the Under Secretary of Defense (Comptroller) created a policy stating that "military members and civilian employees who log onto myPay after a date specified on the homepage will be consenting to receive electronic

Figure 3.3
DFAS Workload Over Time, by Type of Leave-and-Earning Statement, Fiscal Year 2005–2013

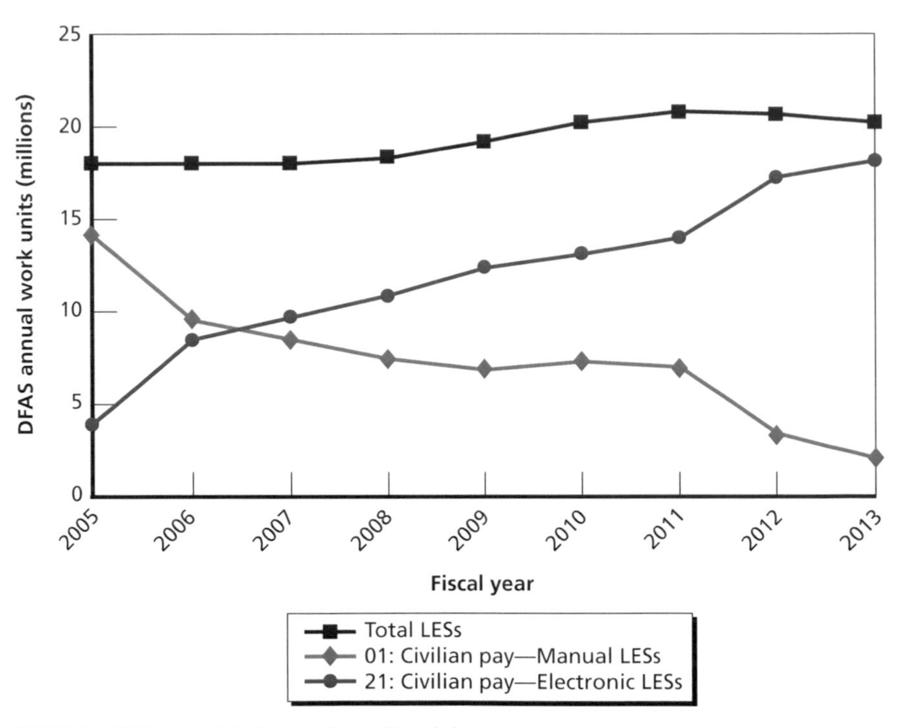

SOURCE: DFAS-provided annual workload data.
RAND *RR866-3.3*

copies of their W-2 and LES, unless they elect to 'turn on' receipts of hard copies by selecting that delivery option on the myPay website" (Chu and Jonas, 2005). On September 30, 2011, the DoD turned off hard copy mailings of LESs to all nonbargaining unit civilians and military members unless the individuals turned on hard copy LES delivery through myPay (Hale, 2011).

Most of DFAS's other outputs do not offer manual-automated price differences, so Figures 3.2 and 3.3 present the cleanest illustrations of the potential effects of offering lower prices for accepting automated approaches.[2]

A further pricing reform that DFAS has implemented in recent years (consistent with a suggestion that we made in Keating et al., 2003) is adding more customer-specific pricing tied to estimates of the cost to serve different customers. As shown in Figure 3.4, DFAS used to charge all customers the same price per active military pay account. Customer-specific prices for this output were introduced in 2003.

When we asked for an explanation about the Marine Corps' lower price, SMEs told us that the Marine Corps Total Force System is superior to the other services' personnel and pay systems and therefore puts less burden on DFAS. Varying prices for a given output by customer rewards low-cost and low-difficulty customers with lower prices while charging more to customers who place greater burdens on DFAS.

Other outputs with military customer–specific prices include commercial payments, reserve military pay, and travel vouchers. Meanwhile, DFAS charges for accounting services, their single most costly output, on a customer-specific basis and per direct billable hour. (Both the number of hours of accounting services billed and the per-hour billing rate vary across DFAS customers.)

While we are supportive of customer-specific pricing, the cost data that we received from DFAS did not include customer-level attribution (i.e., which costs were borne in support of which customers), so we were unable to assess the validity of DFAS's customer-specific prices.

---

[2] There is a sizably reduced price offered for using disburse-only travel voucher services (Output 27) as opposed to manual travel vouchers (Output 7). However, disburse-only vouchers were already predominant in 2005, the first year for which we have workload data.

Figure 3.4
DFAS Prices for Active Military Pay Accounts, by Customer, Fiscal Year 1995–2014

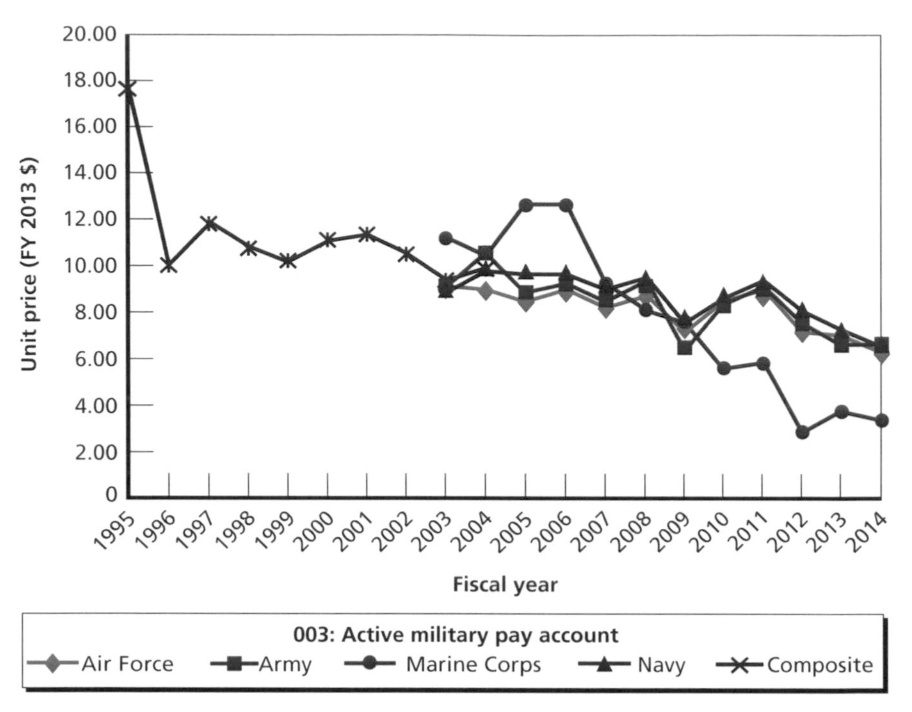

SOURCE: DFAS price lists.
RAND *RR866-3.4*

While Figures 3.2 and 3.3 suggest customer price elasticity of demand (i.e., customers responding to lower prices by evolving toward more-automated approaches), DFAS has some outputs—such as active, reserve, and retired military pay—for which DFAS customer demand is completely price inelastic. The customers can neither use a different provider (or provide the service themselves), nor change their intrinsic demand level in response to DFAS prices. Also, there is not a more-automated version of these outputs that customers might be induced to adopt.

Illustrating this phenomenon, Figure 3.5 shows nearly completely inelastic demand for retired military pay services (i.e., paying military retirees) across the four services. Each horizontal line in Figure 3.5

Figure 3.5
Military Customers' Annual Demands for Retired Military Pay Services, Fiscal Year 2005–2013

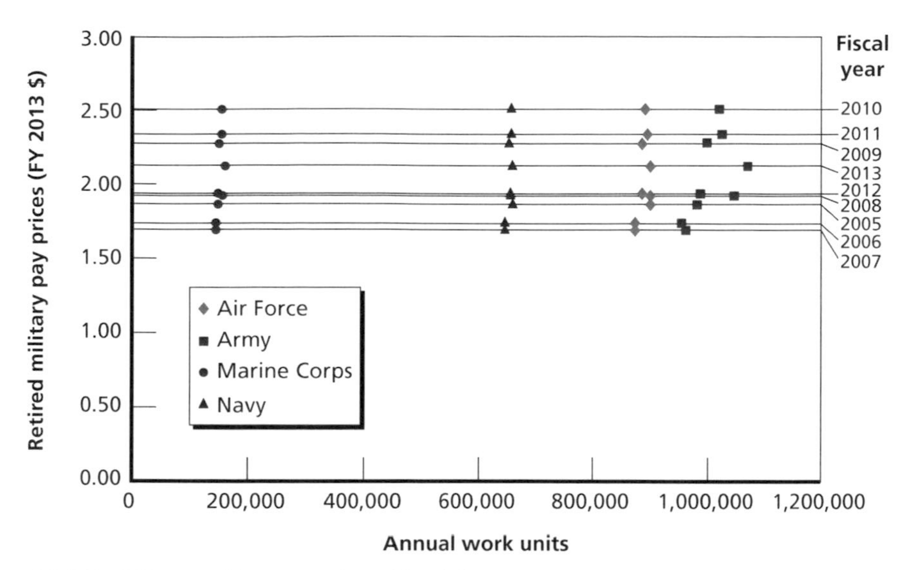

SOURCE: DFAS-provided annual workload data.
RAND *RR866-3.5*

corresponds to a fiscal year's price for that output. (The FY 2008 and FY 2012 prices were almost identical in constant FY 2013 terms—$1.94 and $1.93, respectively.) Unlike many DFAS outputs, retired military pay prices do not vary across customers within a fiscal year.

While the demands shown in Figure 3.5 are clearly highly inelastic, customers may still have some flexibility in the quality and accuracy of the data that they provide to DFAS and, hence, the magnitude of the burden that they place on DFAS. To the extent that different customers impose different burdens on DFAS for retired military pay services, there is an argument in favor of charging them different prices for this output. However, the cost data that we received were insufficient for us to assess the validity of these customer-specific prices. Figure 3.5 provides no evidence of customers having any price elasticity in their quantity of retired military pay services (nor would we logically expect any such price elasticity).

CHAPTER FOUR

Assessing DWCF Pricing in DFAS

This chapter provides an assessment of the advantages and disadvantages of DWCF pricing in DFAS.

Advantages

As documented in Chapter Three, DFAS has had success inducing customers to adopt more-automated approaches, though it is unclear to what extent pricing has driven this favorable outcome.

We also note that DFAS makes considerable use of customer-specific pricing—that is, many DFAS outputs have specific prices for specific military customers. This practice rewards customers for putting less burden per action on DFAS.

DWCF pricing encourages ongoing cost-related dialog, both within DFAS and between DFAS and its customers. Some of that dialog can be negative in tone, but, as one SME noted, customer gripes about prices "were, in some sense, the whole point!" One would not want customers to be oblivious to the costs that their choices impose upon DFAS.

Also, being a DWCF entity provides greater managerial flexibility to DFAS than if it were dependent on direct appropriations. Because it operates within a DWCF, the funding that it receives from its customers is not constrained by the fiscal year. For example, DFAS could continue to operate during sequestration to ensure that military personnel were paid.

On a practical level, DFAS could not take on non-DoD workload without a mechanism to charge non-DoD customers. DFAS was originally created to harvest economies of scale in provision of finance and accounting services. Those same economies of scale argue in favor of DFAS taking on additional non-DoD workload. Non-DoD customers provide an opportunity to spread DFAS overhead across more customers, reducing the burden borne by DoD customers. As shown in Figure 4.1, for many years, non-DoD work has remained a minor portion of DFAS's total revenue (on the order of 5 percent), according to DFAS Working Capital Fund budget submissions. In recent years, the Department of Veterans Affairs has been DFAS's largest non-DoD customer.

We have found no evidence that DoD customers subsidize non-DoD customers—that is, no evidence that DoD customers are made

Figure 4.1
Share of DFAS Revenue from Non-DoD Customers, Fiscal Year 2000–2014

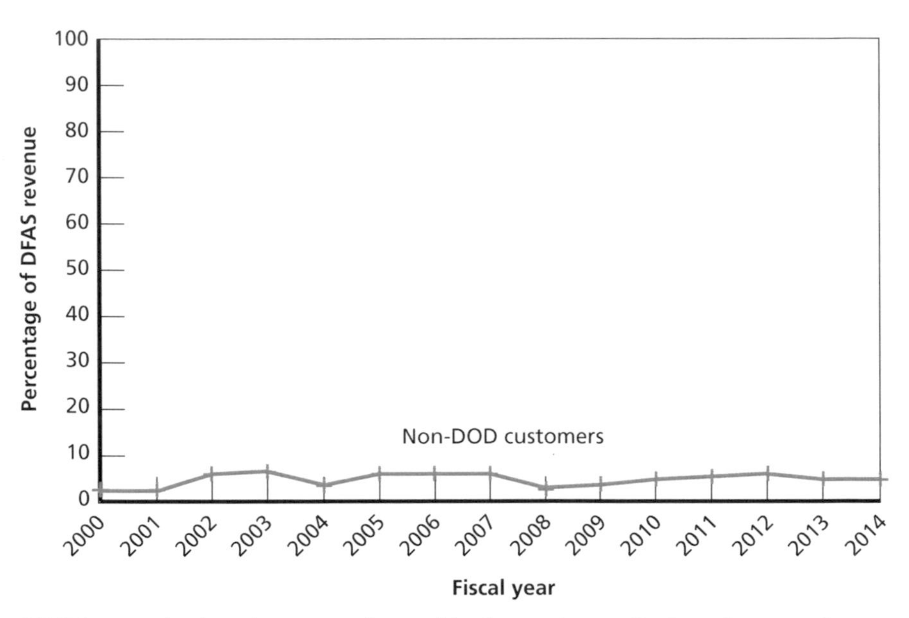

SOURCE: DoD budget documentation and budget estimates (DoD, various years).
RAND *RR866-4.1*

any worse off by DFAS performing work for non-DoD customers. Instead, limited evidence suggests that non-DoD customers more than pay their way, reducing the amount of DFAS overhead that must be covered by DoD customers.

In general, the cost data that we received from DFAS do not allow us to separate the costs of non-DoD customers from those of DoD customers within a given output. An exception to this generalization is Output 51, non-DoD ePayroll civilian pay (which, by definition, only applies to non-DoD customers).

Unfortunately, the cost data for this output are not coherent: Costs only started to be accumulated against this output in 2010, four years after the workload data show the output was sold to non-DoD customers. Further, those costs that have heretofore been attributed to this output represent only a few cents per recorded work unit, vastly below the price that DFAS has charged non-DoD customers for this service. The safest thing that we can say is that we found no evidence of DoD customers subsidizing non-DoD customers for this output.

DFAS has expressed interest in taking on additional non-DoD workload. For example, DFAS's FY 2006 financial report noted that "actions implementing the President's Management Agenda have added new customers from outside the DoD, including the Department of Energy, the Department of Veterans Affairs, the Environmental Protection Agency, and the Department of Health and Human Services" (DFAS, 2006, p. 2).

DFAS efforts to expand non-DoD workload have been impaired, our SMEs told us, by DoD-wide headcount caps applied to DFAS. As these headcount caps have restricted expansion of profitable non-DoD business, DoD customers have ended up bearing greater DFAS overhead costs than would otherwise have been the case.

To the extent that non-DoD work is viewed as being desirable for DFAS, this preference would argue in favor of some sort of pricing mechanism that allows DFAS to generate revenue from those customers.

Disadvantages

There are, however, concerns about DWCF pricing.

The structure may impose greater cost accounting burden than would a direct appropriation mechanism. For example, in order to formulate its prices, DFAS has to attribute its costs to specific products. Such a challenge might not arise if DFAS were wholly funded by direct appropriation. (On the other hand, product-level cost attribution might be necessary for effective management even if DFAS were wholly funded by appropriation.) A nonlinear pricing arrangement, meanwhile, could present a worrisome possibility of yet-greater burden with both a priced and appropriated component. It is unclear how much additional cost accounting burden nonlinear pricing would impose.

We are also concerned that customers may be developing skepticism about the actual consequences of their cost-saving steps. Customers said that they lacked visibility into the effects of changes in DFAS workload and costs on the prices they pay. We heard customers say that DFAS will simply raise rates or prices on other workload even if DFAS provides a lower price on a specific output. One customer opined that it "looks like game playing to us." Consistent with these concerns, an Office of the Inspector General report said that "DFAS had not developed procedures to routinely compare costs and revenues at the output level" (DoD, 2012b, p. 4). DFAS could improve the transparency of its cost structure and pricing determination process.

As is true of other DWCF providers, DFAS's DWCF expected average cost prices are almost certainly above marginal costs because they build in fixed (output-invariant) costs. As a result, DFAS costs will not fall commensurably when workload falls. DFAS's DWCF prices, for that reason, send too strong a signal to customers, who are led to expect greater cost savings from reforms (e.g., adoption of automated approaches) than can reasonably be achieved. Customers' skepticism has increased as they have learned this reality.

We assert that a DFAS DWCF price is not a stand-alone statistic. Instead, any DFAS price is implicitly paired with a workload projection, with the price simply being the ratio of expected costs divided by expected quantity (along with, as discussed in Chapter One, possible adjustments for prior-year profits or losses).

Therefore, changes in quantity should not be expected to change DFAS costs (and, ultimately, customer bills) by the product of the price and the quantity change. DFAS has a small number of customers, so when customers change their quantity demanded, it affects the prices that DFAS charges. By contrast, individual consumers are accustomed to being price-takers, meaning that a change in quantity purchased leads to a proportional change in customer expenditures. This disconnection is a source of recurring disappointment for DFAS customers, because their bills do not fall as much as expected in response to customer cost-saving reforms. The fundamental structure of DWCF expected average cost pricing drives considerable customer disappointment, as we learned in our interviews.

CHAPTER FIVE

Conclusions and Recommendations

As noted in Chapter One, we analyzed two broad categories of criteria by asking two questions:

1. Does the DWCF structure provide appropriate incentives to DFAS?
2. Does the DWCF structure provide appropriate incentives to DFAS's customers?

Incentives to DFAS

Under current DoD policy, DFAS is a monopoly—that is, DoD customers must purchase accounting and finance services from DFAS as opposed to performing the services themselves or purchasing them from an outside provider, such as private-sector firms (e.g., ADP) or other government providers (e.g., the Department of Agriculture's National Finance Center).[1] (However, the introduction of enterprise resource planning systems may create opportunities for customers to

[1] Some DFAS employees have, however, been subject to A-76 competitions, in which private-sector firms competed with DFAS employees to perform certain types of workload. In all but one of these competitions, the government employees retained the work. The exception was Military Retired and Annuitant Pay Services, which was contracted out to Affiliated Computer Services. However, this workload was brought back in-house in FY 2009. There was considerable controversy concerning this award discussed in an Office of the Inspector General report on public-private competition for DFAS (DoD, 2003b). Gates and Robbert (2000) provide a more general discussion of competitive sourcing in the DoD.

take over some workload that is currently performed by DFAS.[2]) As a consequence, DFAS may lack robust incentives to provide its services in a cost-effective way.

The DFAS total cost trends shown in Chapter Two are heartening, however. DFAS's real costs have trended downward over a lengthy period during which DoD's overall manning and budget have grown. Customer and DoD oversight pressure, as well as the intrinsic motivation of DFAS leaders, have sufficed, at least heretofore, to achieve a favorable cost outcome.

Funding DFAS with direct appropriations would be an alternative approach. If it were funded this way, DFAS could be told exactly the budget level at which it must operate. However, DFAS might then be unable to fulfill its mission satisfactorily. By contrast, as noted, the current DWCF approach engenders an ongoing cost-related dialog between DFAS and its customers, although our SME interviews suggest that this dialog has not always resulted in the customer being pleased with the prices that DFAS charges for the outputs it provides. Nevertheless, it might be preferable to have customer-DFAS dialog decide which DFAS outputs can be reduced rather than leaving that decision in the hands of appropriators.

Incentives to DFAS Customers

DWCF prices provide considerable incentives to DFAS customers. Customers are rewarded (at least in the short run) for adopting approaches that put less burden on DFAS. To the extent that genuine DFAS cost savings accrue from customer changes, we ultimately expect a diminution in customers' bills (as opposed to simply reallocating DFAS overhead across outputs). Indeed, returning to Chapter Four's discussion,

---

[2] The military services are in various stages of adopting enterprise resource planning systems for finance and accounting, such as the Army's General Fund Enterprise Business System, the Navy Enterprise Resource Planning System, and the Air Force's Defense Enterprise Accounting and Management System. When these systems are fully operational, they should be capable of performing some of the same financial transactions and accounting reconciliations currently performed by DFAS.

DWCF prices provide overly large incentives for customers to change their behavior. Actual DFAS costs will not fall as much as the DWCF price differences customers observe. DFAS prices have two basic components: DFAS's variable cost of providing the service and a constant percentage markup to cover DFAS fixed costs. Therefore, the observed price difference between two approaches is composed of the difference in variable costs multiplied by a markup. The observed price difference between two approaches is therefore greater than the difference in the variable costs of the two approaches.

By contrast, if DFAS were funded by direct appropriation, only moral suasion or external mandates might be used to encourage customers to take cost-saving steps, such as adopting automated outputs and taking other steps to minimize their burden on DFAS.

We conclude that DWCF prices provide more incentives to DFAS customers than to DFAS itself.

Possible Competition for DFAS

An aggressive reform, proposed by several of our customer SMEs, would be to change DoD policy to allow other governmental or private-sector providers to compete with DFAS for DoD business. DoD policy could also explicitly allow the services themselves to create internal, within-service competition with DFAS.

DWCF pricing rules have caused problems when customers have had alternatives. For example, Brauner et al. (2000) noted how the U.S. Army's Forces Command (FORSCOM) set up an intracommand redistribution and repair system to reduce the amount of workload it sent to the Army's then-underutilized depot repair system. FORSCOM customers saved operations and maintenance funds for other uses by not buying as many services from the Army's depots. However, these were not necessarily savings from an Army-wide perspective, because there was a sizable discrepancy between Army depot prices and actual variable costs. In the presence of fixed costs and expected average cost pricing, total costs to DoD may go up, not down, when customers shift workload away from DWCF providers.

If policy were changed and DFAS faced competition, its current pricing structure would be inadequate because it does not properly fit with having considerable fixed costs.

An assessment of the overall desirability of allowing competition with DFAS was beyond the scope of this analysis. Certainly, introducing competition would substantially change DFAS's incentives and force a change in how it prices its services.

Concluding Remarks

A DWCF provider's prices offer incentives to its customers. Customers can and do respond to these incentives, so it is important that these incentives induce desirable behavior. In DFAS's case, it appears that its reduced prices for adopting automated approaches have worked as intended.

DWCF prices do not, however, provide efficiency incentives to the DWCF provider itself. One way to incentivize a DWCF provider such as DFAS would be allowing competition for services. However, the current price system would not be appropriate under competition because, as in the FORSCOM example, competition in the presence of an unsuitable pricing structure could prove to be more costly to the DoD than no competition at all.

A return to complete reliance on direct appropriations is probably an overreaction to DWCF pricing problems. We would urge, instead, reforming prices (e.g., nonlinear pricing) rather than eliminating them. We would also urge DFAS to provide greater visibility into its cost structures, such as indirect cost allocation and customer-specific costs.

Nonlinear pricing could be implemented in a variety of ways, including

- Quantity discounts, with the DWCF provider charging less per unit if it receives more workload.
- The DWCF provider receiving appropriations to cover its fixed costs, with customer prices equal to the incremental costs of the services that they purchase. A customer would then face marginal

cost, not a multiple above marginal cost, as its output price. If the customer left the DWCF provider, provider cost savings would equal marginal cost, so the customer receives the correct signal from the price system.

As mentioned in Chapter One, DoD Financial Management Regulation and 10 U.S.C. 2208 require DWCF entities to set their prices based on full cost recovery. Further examination of whether and how nonlinear pricing can be accommodated within the full cost recovery mandate would be useful.

Nonlinear pricing would be new in DFAS. We urge experimentation with pilot projects to better understand such an approach's implementation challenges, costs, and ultimate consequences.

APPENDIX

RAND Subject-Matter Expert Questions

RAND developed questions that we used to guide our conversations with DFAS customers and DFAS leaders. In many cases, we sent these questions ahead of time to our interviewees. In this appendix, we present those questions.

RAND Questions for DFAS Customers

RAND's National Defense Research Institute is working on a project entitled Defense Working Capital Fund–Related Insights from Analysis of the Defense Finance and Accounting Service, sponsored by the Office of the Secretary of Defense for Cost Assessment and Program Evaluation (OSD-CAPE). The project seeks to understand the advantages and disadvantages of funding the Defense Finance and Accounting Service (DFAS) through a Defense Working Capital Fund (DWCF) versus using direct appropriations.

As part of this project, the RAND research team seeks to understand the views and perceptions of DFAS customers. We do not, however, intend to attribute any specific views to any specific DFAS customers, either in RAND communication with OSD-CAPE or in any written RAND report.

We therefore ask the following questions of DFAS customer representatives:

- What is the nature and frequency of your interaction with DFAS?

- What are representative challenges/issues you face concerning DFAS?
- What do you see as DFAS's greatest strengths? Biggest areas of concern?
- Does being a DWCF entity benefit DFAS? If so, how?
- What is the current process by which DFAS negotiates workload and rates with its customers? Has that process evolved over time?
- Does your organization respond to DFAS price levels and/or changes? If so, how?
- Has your organization had experience with transitioning from manual to automated DFAS outputs? Did such transitions progress successfully, in your opinion? Did DFAS being a DWCF entity abet, retard, or have no meaningful effect on the success of those efforts?
- Have you ever compared the prices you pay for customer-differentiated outputs (e.g., active military pay, reserve military pay, commercial payments, travel disbursements) to those paid by other DFAS customers? Do you feel that customer-differentiated prices are determined fairly? Has DFAS ever discussed with you how they calculate customer-differentiated prices?
- If you could propose one major reform to how DFAS operates, what reform would you propose? Do you have insight as to why your preferred reform hasn't heretofore been implemented?
- Do you feel that DFAS pricing might work better (and/or be fairer or otherwise more desirable) if DFAS were allowed to use nonlinear pricing (e.g., an "open the door" fee, quantity discounts)?
- What might be the advantages and disadvantages of shifting toward appropriated funding for some or all of DFAS's workload?

RAND Questions for DFAS Leaders

RAND's National Defense Research Institute is working on a project entitled Defense Working Capital Fund–Related Insights from Analysis of the Defense Finance and Accounting Service, sponsored by the Office of the Secretary of Defense for Cost Assessment and Program Evaluation (OSD-CAPE). The project seeks to understand the advantages and disadvantages of funding the Defense Finance and Accounting Service (DFAS) through a Defense Working Capital Fund (DWCF) versus using direct appropriations. The project will also assess whether insights derived from DFAS may be more broadly applicable to other DoD DWCFs.

While we seek subject matter expert judgments about DFAS, we will not specifically attribute any remarks in presentations to OSD-CAPE or in any RAND publication. Put differently, while we will acknowledge folks who were kind enough to speak with us, we will not tie their remarks to them by name.

We therefore ask the following questions of DFAS leaders:

- What are your responsibilities within DFAS? How long have you worked for DFAS? What other positions have you held within DFAS? What prior experiences did you have before you came to work for DFAS?
- What are typical issues or challenges you face in your job?
- Do you interact much with DFAS customers as part of your job? If so, what is the nature and frequency of those interactions? What type/level of customers do you interact with?
- Do you participate in DFAS's price-setting/negotiation process? To your knowledge, has that process evolved over time?
- In your experience, do DFAS customers respond to DFAS price levels and/or changes? If so, how?
- Have DFAS efforts to transition customers to less manual, more-automated (e.g., web-based) approaches progressed as successfully and quickly as you envisioned? Did DFAS being a DWCF entity abet, retard, or have no meaningful effect on the success of those efforts?
- How have the services' adoption of enterprise resource planning (ERP) systems changed their interactions with DFAS? Do you expect further changes because of ERP?
- What do you see as DFAS's greatest strengths? Biggest areas of concern?

- Does being a DWCF entity benefit DFAS? If so, how?
- If you could propose one major reform to how DFAS operates, what reform would you propose? Do you have insight as to why your preferred reform hasn't heretofore been implemented?
- Would you look favorably or unfavorably on DFAS increasing the non-DoD share of its workload?
- Do you feel that DFAS pricing might work better (and/or be fairer or otherwise more desirable) if DFAS were allowed to use nonlinear pricing (e.g., an "open the door" fee, quantity discounts)?
- What might be the advantages and disadvantages of shifting toward appropriated funding for some or all of DFAS's workload?

References

Brauner, Marygail K., Ellen M. Pint, John R. Bondanella, Daniel A. Relles, and Paul Steinberg, *Dollars and Sense: A Process Improvement Approach to Logistics Financial Management*, Santa Monica, Calif.: RAND Corporation, MR-1131-A, 2000. As of December 18, 2014:
http://www.rand.org/pubs/monograph_reports/MR1131.html

Byrnes, Patricia E., "Defense Business Operating Fund: Description and Implementation Issues," *Public Budgeting & Finance*, Winter 1993.

Chu, David S. C., and Tina W. Jonas, "Policy for Electronic Wage and Tax Statements and Leave and Earning Statements Through myPay" Under Secretary of Defense for Personnel and Readiness and Under Secretary of Defense (Comptroller) memorandum, April 29, 2005.

Defense Finance and Accounting Service, *Working Capital Fund: Fiscal Year 2006 Financial Report*, November 2006.

———, *Working Capital Fund: Fiscal Year 2007 Financial Report*, November 2007.

———, "DFAS Locations," web page, April 1, 2011. As of June 30, 2014:
http://www.dfas.mil/careers/locations.html

———, "Agency Overview," web page, January 21, 2014. As of August 3, 2014:
http://www.dfas.mil/pressroom/aboutdfas.html

DFAS—*See* Defense Finance and Accounting Service.

DoD—*See* U.S. Department of Defense.

Federal Register, "Mandatory Use of Wide Area WorkFlow," Defense Federal Acquisition Regulation Supplement Case 2006-D049, Vol. 72, No. 156, August 14, 2007. As of August 13, 2014:
http://www.acq.osd.mil/dpap/dars/dfars/changenotice/2007/20070814/E7-15928.htm

———, "Electronic Submission of Payment Requests and Receiving Reports," Defense Federal Acquisition Regulation Supplement, 252.232-7003, January 2012. As of December 17, 2014:
http://www.acq.osd.mil/dpap/dars/dfars/html/current/252232.htm#252.232-7003

Friend, Gregory C., *An Examination of the Stabilized Rate Setting Process within the Defense Business Operations Fund*, Monterey, Calif.: Naval Postgraduate School thesis, June 1995.

Gates, Susan M., and Albert A. Robbert, *Personnel Savings in Competitively Sourced DoD Activities: Are They Real? Will They Last?* Santa Monica, Calif.: RAND Corporation, MR-1117-OSD, 2000. As of December 18, 2014:
http://www.rand.org/pubs/monograph_reports/MR1117.html

Hale, Robert F., "Electronic Leave and Earning Statement (eLES)" Department of the Navy memorandum, August 9, 2011. As of August 14, 2014:
http://www.public.navy.mil/ia/Documents/ASN_FMC_MRA_eLES_JntLtr.pdf

Keating, Edward G., "RAND Research Suggests Changes in Department of Defense Internal Pricing," Santa Monica, Calif.: RAND Corporation, IP-216-DFAS, 2001. As of December 18, 2014:
http://www.rand.org/pubs/issue_papers/IP216.html

Keating, Edward G., and Susan M. Gates, *Defense Working Capital Fund Pricing Policies: Insights from the Defense Finance and Accounting Service*, Santa Monica, Calif.: RAND Corporation, MR-1066-DFAS, 1999. As of December 18, 2014:
http://www.rand.org/pubs/monograph_reports/MR1066.html

Keating, Edward G., Susan M. Gates, Jennifer E. Pace, Christopher Paul, and Michael G. Alles, *Improving the Defense Finance and Accounting Service's Interactions with Its Customers*, Santa Monica, Calif.: RAND Corporation, MR-1261-DFAS, 2001. As of December 18, 2014:
http://www.rand.org/pubs/monograph_reports/MR1261.html

Keating, Edward G., Susan M. Gates, Christopher Paul, Aimee Bower, Leah Brooks, and Jennifer E. Pace, *Challenges in Defense Working Capital Fund Pricing: Analysis of the Defense Finance and Accounting Service*, Santa Monica, Calif.: RAND Corporation, MR-1597-DFAS, 2003. As of December 18, 2014:
http://www.rand.org/pubs/monograph_reports/MR1597.html

Office of Personnel Management, "Data, Analysis & Documentation: Federal Employment Reports, Historical Federal Workforce Tables, Executive Branch Civilian Employment Since 1940," undated a. As of August 1, 2014:
http://www.opm.gov/policy-data-oversight/data-analysis-documentation/federal-employment-reports/historical-tables/executive-branch-civilian-employment-since-1940/

———, "Data, Analysis & Documentation: Federal Employment Reports, Historical Federal Workforce Tables, Total Government Employment Since 1962," undated b. As of August 1, 2014:
http://www.opm.gov/policy-data-oversight/data-analysis-documentation/federal-employment-reports/historical-tables/total-government-employment-since-1962/

U.S. Department of Defense, "Defense Finance and Accounting Service: Overview," *Defense Wide Budget Documentation—FY2002: Defense Working Capital Fund*, June 2001. As of December 18, 2014:
http://comptroller.defense.gov/Portals/45/Documents/defbudget/fy2002/budget_justification/pdfs/06_Defense_Working_Capital_Fund/dla_dfas.pdf

———, "Defense Finance and Accounting Service: Overview," *Defense Wide Budget Documentation—FY2003: Defense Working Capital Fund*, February 2002. As of December 18, 2014:
http://comptroller.defense.gov/Portals/45/Documents/defbudget/fy2003/budget_justification/pdfs/06_Defense_Working_Capital_Fund/DFAS.pdf

———, "Fiscal Year (FY) 2004/FY 2005 Biennial Budget Estimates: Defense Finance and Accounting Service (DFAS)," *Defense Wide Budget Documentation—FY2004: Defense Working Capital Fund*, February 2003a. As of December 18, 2014:
http://comptroller.defense.gov/Portals/45/Documents/defbudget/fy2004/budget_justification/pdfs/01_Operation_and_Maintenance/Volume_1_-_DW_Justification/DFAS_FY04-05_PB.pdf

———, *Infrastructure and Environment: Public/Private Competition for the Defense Finance and Accounting Service Military Retired and Annuitant Pay Functions*, Office of the Inspector General, D-2003-056, March 21, 2003b. As of August 2, 2014:
http://www.dodig.mil/audit/reports/fy03/03-056.pdf

———, *Defense Working Capital Fund Defense-Wide Fiscal Year (FY) FY 2005 Budget Estimates: Operating and Capital Budgets*, February 2004. As of December 18, 2014:
http://comptroller.defense.gov/Portals/45/documents/defbudget/fy2005/budget_justification/pdfs/06_Defense_Working_Capital_Fund/DWCF-_DLA(co)DFAS(co)DISA(co)_and_DSS.pdf

———, *Defense Working Capital Fund Defense-Wide Fiscal Year (FY) FY 2006/2007 Budget Estimates: Operating and Capital Budgets*, February 2005. As of December 18, 2014:
http://comptroller.defense.gov/Portals/45/Documents/defbudget/fy2006/budget_justification/pdfs//06_Defense_Working_Capital_Fund/DWWCF_PB_06_Operating.pdf

———, *Defense Working Capital Fund Defense-Wide Fiscal Year (FY) FY 2007 Budget Estimates: Operating and Capital Budgets*, February 2006. As of December 18, 2014:
http://comptroller.defense.gov/Portals/45/Documents/defbudget/fy2007/budget_justification/pdfs/06_Defense_Working_Capital_Fund/DWWCF_%20-_Operating.pdf

———, *Defense Working Capital Fund Defense-Wide Fiscal Year (FY) FY 2008/2009 Budget Estimates: Operating and Capital Budgets*, February 2007. As of December 18, 2014:
http://comptroller.defense.gov/Portals/45/Documents/defbudget/fy2008/budget_justification/pdfs/06_Defense_Working_Capital_Fund/DW-WCF_%20-_Operating.pdf

———, *Defense Working Capital Fund Defense-Wide Fiscal Year (FY) FY 2009 Budget Estimates: Operating and Capital Budgets*, February 2008. As of December 18, 2014:
http://comptroller.defense.gov/Portals/45/Documents/defbudget/fy2009/budget_justification/pdfs/06_Defense_Working_Capital_Fund/DW-WCF_%20-_Operating.pdf

———, *Defense Working Capital Fund Defense-Wide Fiscal Year (FY) FY 2010 Budget Estimates: Operating and Capital Budgets*, May 2009. As of December 18, 2014:
http://comptroller.defense.gov/Portals/45/Documents/defbudget/fy2010/budget_justification/pdfs/06_Defense_Working_Capital_Fund/PB10_DW-WCF-Operating.pdf

———, *Defense Working Capital Fund Defense-Wide Fiscal Year (FY) FY 2011 Budget Estimates: Operating and Capital Budgets*, February 2010a. As of December 18, 2014:
http://comptroller.defense.gov/Portals/45/Documents/defbudget/fy2011/budget_justification/pdfs/15_Revolving_funds/02-RF_Operating_Budget_Estimates_fy2011.pdf

———, "Budget Formulation and Presentation (Chapters 4–19)," Financial Management Regulation, DoD 7000.14-R, Vol. 2B, June 2010b. As of October 2, 2014:
http://comptroller.defense.gov/Portals/45/documents/fmr/Volume_02b.pdf

———, *Defense Working Capital Fund Defense-Wide Fiscal Year (FY) FY 2012 Budget Estimates: Operating and Capital Budgets*, February 2011. As of December 18, 2014:
http://comptroller.defense.gov/Portals/45/Documents/defbudget/fy2012/budget_justification/pdfs/06_Defense_Working_Capital_Fund/PB_12_DWWCF_Operating_Budget.pdf

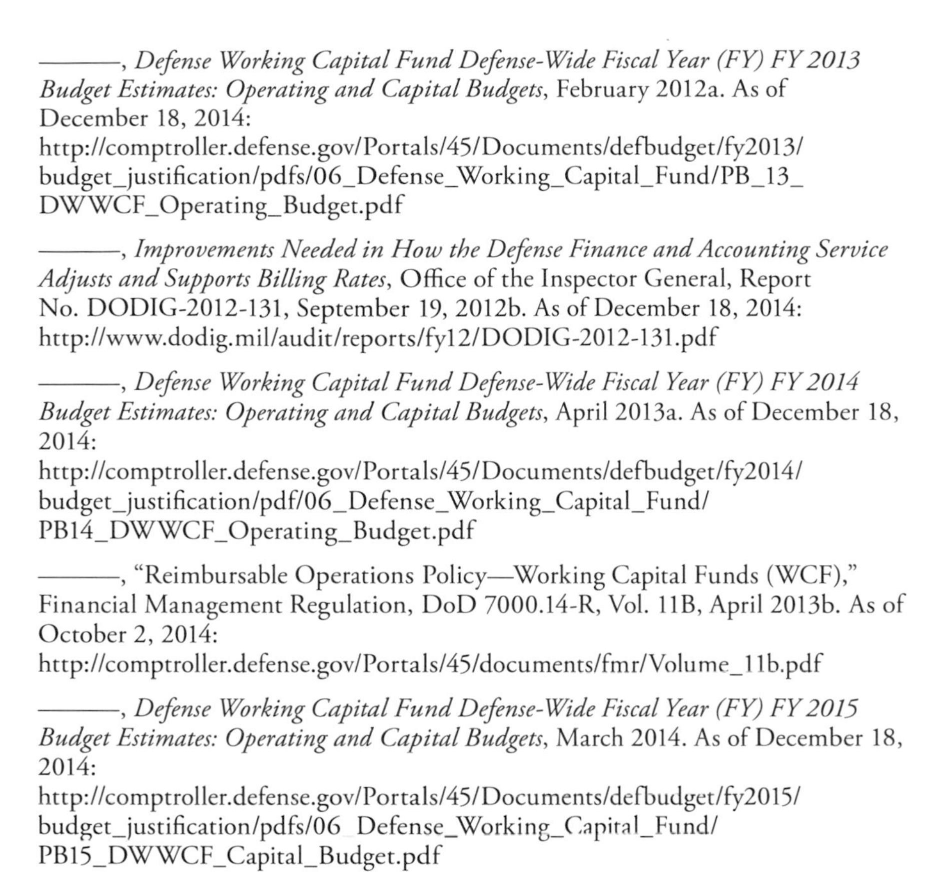

———, *Defense Working Capital Fund Defense-Wide Fiscal Year (FY) FY 2013 Budget Estimates: Operating and Capital Budgets*, February 2012a. As of December 18, 2014:
http://comptroller.defense.gov/Portals/45/Documents/defbudget/fy2013/budget_justification/pdfs/06_Defense_Working_Capital_Fund/PB_13_DWWCF_Operating_Budget.pdf

———, *Improvements Needed in How the Defense Finance and Accounting Service Adjusts and Supports Billing Rates*, Office of the Inspector General, Report No. DODIG-2012-131, September 19, 2012b. As of December 18, 2014:
http://www.dodig.mil/audit/reports/fy12/DODIG-2012-131.pdf

———, *Defense Working Capital Fund Defense-Wide Fiscal Year (FY) FY 2014 Budget Estimates: Operating and Capital Budgets*, April 2013a. As of December 18, 2014:
http://comptroller.defense.gov/Portals/45/Documents/defbudget/fy2014/budget_justification/pdf/06_Defense_Working_Capital_Fund/PB14_DWWCF_Operating_Budget.pdf

———, "Reimbursable Operations Policy—Working Capital Funds (WCF)," Financial Management Regulation, DoD 7000.14-R, Vol. 11B, April 2013b. As of October 2, 2014:
http://comptroller.defense.gov/Portals/45/documents/fmr/Volume_11b.pdf

———, *Defense Working Capital Fund Defense-Wide Fiscal Year (FY) FY 2015 Budget Estimates: Operating and Capital Budgets*, March 2014. As of December 18, 2014:
http://comptroller.defense.gov/Portals/45/Documents/defbudget/fy2015/budget_justification/pdfs/06_Defense_Working_Capital_Fund/PB15_DWWCF_Capital_Budget.pdf

U.S. General Accounting Office, *Foreign Military Sales: DOD's Stabilized Rate Can Recover Full Cost*, GAO/AIMD-97-134, September 1997. As of December 18, 2014:
http://www.gao.gov/assets/230/224652.pdf